*A Stone's Throw; Two Islands Beyond the Peninsula*
ISBN-13: 978-1-939722-08-9
ISBN-10: 1-939722-08-X
https://www.createspace.com/6336867

ACKNOWLEDGEMENTS
Special thanks goes to Christine Sjogren, who continues to provide moral support, painstaking editing, and valuable constructive suggestions for this book as well as for the first four BioFables books. Thanks to Kristin Aufmann for sharing her personal knowledge derived from many visits to Washington and Rock Islands along with the "before" and "after" photos of a fish boil's *overboil* (page 34 and the front cover) and for introducing the photographic artistry of Fingerstyle guitarist and songwriter Daryl Shawn (http://www.darylshawn.com/), who provided the photo of the rocky shoreline of Schoolhouse Beach. United States maps are from The National Atlas.

Published by
**Technology Management Associates, Inc.**
1699 Wall Street, Suite 515
Mount Prospect, Illinois 60056

Previous books in *BioFables, Series 1*

Prequels: for readers ages 10 and higher
Book 1: *Whoosh; Old Faithful Uncovers a Mystery* https://www.createspace.com/5363808
Book 2: *E-I-E-I-Uh oh; Down on the (Family) Farm* https://www.createspace.com/5688511

For readers 7-12
Book 3: *Sand Sack; Singing Sands Sing No Secrets* https://www.createspace.com/5917670
Book 4: *Palisades Escapades; Bats and a Bird Bring Bumps and Bruises*
https://www.createspace.com/6115848

# A Stone's Throw

BioFables

## Two Islands Beyond the Peninsula

Joanne Gucwa

**Series 1**
**The Adventures Begin**

Book 5

Published by **Technology Management Associates, Inc.**, Mount Prospect, IL USA

# CHAPTERS

# HELPFUL HINTS

## About the Two Islands

Most people who visit Door County never get beyond Door County's peninsula itself. This story is different, because most of the adventures take place on Washington Island and Rock Island. You take one ferry to get to Washington Island from the Door County Peninsula, and a different ferry to get to Rock Island from Washington Island.

During their last adventure, Mallory and Melody Maloney learned about glaciers and how those long-ago ice formations created sand dunes and other features of the land. On this trip they visit a beach that has no sand; they watch the sun set and then rise the next morning over water while standing at the same place on land; and they experience the flare-up of flames during a traditional fish boil, ...to name just a few of their adventures.

As are all BioFables stories, this book is a work of fiction. While real place names are used and the science is also real, all people and incidents are invented.

## About the BioFables Series

All *BioFables* books have spaces in the outside margins for you to write notes for yourself. There is also an Index at the back of the book so you can find a section you would like to re-read.

You may find interesting science or other topics that you would like to know more about. We've set up the *BioFables* website (www.biofables.com) where you can find links to online articles on specific topics for each book. Just click the *Series 1* tab and then click on the title *A Stone's Throw* for Book 5. You will see that the links are arranged by chapter. If you don't find the topic you're looking for, it's possible that the topic was already covered in a previous book in this series. Click on earlier book titles to see if you can find the information you want.

Each book in this *The Adventures Begin* series shows you a map so you will know where Melody's and Mallory's adventures are taking them. The star shows where they live (a suburb of Chicago near O'Hare Airport); the arrow pointing to a numbered circle shows the location of their adventure (in this book, it is number 5).

# HAPPY READING!

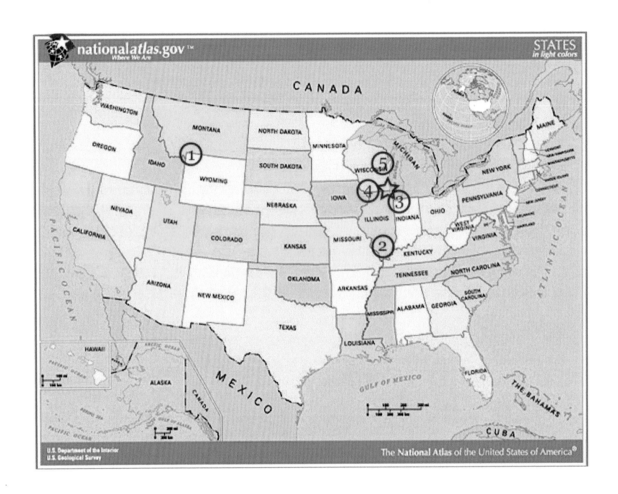

# Chapter 1
## Where Is the Door in Door County?

"We're now passing *Porte des Morts*," the captain of the ferry announced over the speaker system. "For those of you whose French is a little rusty, *Porte des Morts* means *Death's Door*; that is how Door County got its name."

Those were exciting words to Mallory. *Death's Door*, he thought. He walked around all sides of the ferry with Rufus, the family dog. Looking down at the water, he was trying to find a door floating in the water.

"Where's Death's Door, Dad? I want to see what Death's Door looks like."

Mallory's Grandfather smiled and said, "Mal, I think you're looking at Death's Door. See all the water around us? This is the area where many ships have sunk in the past. French explorers learned first-hand how dangerous a narrow passage of water can be. Storms and high winds whip up the waves higher in the passage than they'd be in the open waters of Green Bay, on the left, or Lake Michigan, on the right. Their currents also crash into each other here."

Mallory's twin sister, Melody, said, "You know a lot, Grandpa. That's neat! I think islands are neat."

"How do you know they're neat if you've never been on one?" her brother asked.

Their Mom answered him, "Mal, I'm much more than seven years old, and I can't remember if I've ever been on an island or not. I agree with Mel, though; I think land surrounded by water is neat."

"You're just fond of the water, Ag," their Dad said. "The shoreline isn't far away when you're on a small island like the one we're heading toward, not far away, like in the Colorado mountains where you grew up."

## Two Islands of Door County

This outing was going to be longer than the 2-3 day monthly trips that Grandpa Mike gave as a birthday present when Mallory and Melody reached their seventh birthdays earlier in the year. His intent was to give the twins (and their parents) a chance to enjoy nature's beauty while exploring and learning about some of its secrets.

The Maloney family had decided to enjoy Door County's less-crowded nearby islands: Washington Island and Rock Island.

According to the family's research, the two islands offered lots of new experiences and chances for making new discoveries. An extra bonus: they could see lots more stars than on the crowded mainland.

Everyone got back in the car when the ferry reached the dock at Washington Island. The twin's Dad drove off the ramp and headed toward the inn where they would be staying. There would be no camping this trip. Grandpa Mike said his "old bones" much preferred a bed, even if an air mattress came between him and the hard ground.

## The Many Flavors of Perch

The owners of the inn, Tom, and his wife, Karen, greeted the Maloney family.

"Good evening. You must be the Maloney family. My name is Tom and this is my wife, Karen."

"Hi, Mr. Tom," Melody said. "I'm Melody. People call me Mel. This is my twin brother, Mallory. People call him Mal."

"Hello, Tom and Karen. This is my husband Mort and my father-in-law Mike. I'm Agnes. We're happy to be here, especially since you allow pets in the rooms. Our dog Rufus is certainly happy to be able to come with us."

"We welcome Rufus as well," said Ms. Karen. "You'll find that all of Door County is very dog-friendly."

"We're happy to hear that, Karen. Might we be able to have dinner at your restaurant as soon as we're settled in? We've had a long drive up from the Chicago area, and I think the children are hungry," Grandpa Mike said. Melody and Mallory nodded energetically.

Grandpa Mike was probably as hungry as his grandchildren were.

"Yes, of course," Ms. Karen said. "We'll have a table waiting for you whenever you're ready."

When the family arrived about fifteen minutes later, Mr. Tom led them to a table by a window.

"Oh, look! The water sparkles like glitter," Melody said as she pointed out the window.

"That's Green Bay. We're lucky to see it at sunset," said her Mom.

"It feels like a really big dining room instead of a restaurant," Mallory said as he looked around.

Most of the tables were empty, as it was already quite late. The big room looked even bigger, with only a few people at two tables who were still finishing their meals.

Mr. Tom handed everyone a menu. "Fish is always a good choice. Perch is tonight's special. It was still swimming this morning. Children, would you like milk or juice?"

"Milk, please, Mr. Tom," Mallory and his sister Melody said at the same time.

"How is the perch prepared?" Grandpa Mike asked.

"Take your pick," Ms. Karen said, as she came to their table. "Our cook can steam, bake, broil, fry, sauté, with or without batter, and season it to your liking. But better save room for dessert – our local cherries are legendary."

Like many American seven-year-olds, Melody and Mallory aren't all that fond of fish, and said so.

Their Mom told them, "Food so fresh it was swimming this morning will be a special treat. You need to try it."

Like many food chemists, their Dad tends to look at some meals as an extension of his work. "Let's order five different styles so we can 'share and compare; flavors," their Dad suggested.

"Great idea, Mort," their Mom said. As a wellness instructor, getting her family to eat healthy is also an extension of her work.

Everyone placed their orders, each selecting a different kind of perch dinner. While they were waiting for their salads, Mallory and Melody went to the brochure display at the entrance to the dining room and started collecting some of the maps and brochures of

Washington Island, Rock Island and the Karfi ferry that runs between the two islands.

Mr. Tom brought the salads to the Maloney table. "How long are you were staying on Washington Island?" he asked.

"Our schedule is fairly flexible, Tom," Grandpa Mike said. "We'll probably enjoy four or five days here and a day at Rock Island before heading home. That should give us enough time to visit a few interesting places along Lake Michigan on the way back."

The twins' Mom said, "After doing some research, I thought exploring Washington and Rock Islands would be fun and educational for our children. Our main purpose is to introduce them to the interesting natural wonders of this area. My husband, my father-in-law and I are interested in the county's famous fish boils. The people at the visitors' center on the mainland gave us some ideas, but we'd appreciate your insights as a local resident."


### A Beach with no Sand?

Mr. Tom said, "Schoolhouse Beach is a fascinating place, and it is a genuine natural wonder. It's one of only five beaches in the whole world that doesn't have sand."

Melody frowned, thinking, *how can it be a beach if it doesn't have any sand?* Her brother was confused, too.

Mr. Tom looked at the children and smiled. "You're probably thinking that you can't make sand castles at this beach, right? Well, you won't be disappointed. At this beach, you can build *rock* castles instead. The rocks are so smooth and soft that you can make all sorts of creations that you would never have been able to make with sand."

Mallory perked up right away. "Wow, that sounds really neat!"

"Except when Rufus knocks them down," Melody said, giggling. She remembered how Rufus trampled all over their sand castles at the Indiana Dunes.

"You can also try to see how many times the rocks will skip when you aim them just right over the water," said Mr. Tom.

"That sounds like a lot of fun," Grandpa Mike said. "We could see if Rufus chases them."

"Are the rocks made of limestone?" her Dad asked. "We saw the limestone cliffs at Mississippi Palisades recently."

"Yes, the rocks are made up of limestone and dolomite. Geologists call it dolostone," Mr. Tom said. "These rocks have been polished by glaciers, wind and the water that's constantly washing over them. The rocks don't turn into sand like at regular beaches, though. They're too hard."

*How could rocks be hard and soft at the same time?* Melody wondered.

Mallory wasn't thinking about Mr. Tom's geology lesson. He thought the smooth rocks would make a nice addition to his "collection" of a single BIG rock.

"I do have to give you a warning, though, before you get any ideas of taking some home as souvenirs," Mr. Tom said, guessing what Mallory might be thinking. "In a word, DON'T! Schoolhouse Beach is now a protected area, and there is a large fine if you're caught."

*Uh, oh*, Mallory thought. *I hope nobody says anything about the rock I took from the Indiana Dunes.*

Nobody did.

"That sounds like a fine place to start our day tomorrow," said Mallory's Mom. "How far is it from here?"

"Nothing's very far here on Washington Island. Let me highlight the route for you on this map," Mr. Tom said, as he took out a yellow highlighter and opened one of the brochures the twins had collected. "You shouldn't plan on swimming there with your young ones, though. It gets deep pretty quickly," he warned.

"Thanks for letting us know, Tom," said Mallory's Dad. "From the sound of it, we should have plenty of fun on shore."

Grandpa Mike then asked, "Tom, how many miles wide is Washington Island, and how many miles long?"

Before Mr. Tom could answer, the twins' Mom said, "Tom, we'd love to learn more about Washington Island from you and your wife. If you're not too busy, we'd love to have you join us as we enjoy your wonderful perch and hear more about this area, wouldn't we?" she looked at her family. All four heads nodded in agreement.

Mr. Tom had chosen the largest table in his restaurant for the Maloney family. He and his wife usually ate their dinner in the kitchen after all their other guests had left. He secretly hoped that this evening might be a bit more interesting.

"That would be nice. Thank you. I'll check on your dinners and be back in a bit," said Mr. Tom.

Just then Ms. Karen came out of the kitchen with a tray of five different perch dinners. Mr. Tom and his wife placed each plate on the table in front of the person who had ordered that particular dish.

The twins' Mom repeated her invitation for the hosts to join her family.

"It's very kind of you. Thank you. We haven't eaten yet, since we usually eat after our last guest is finished. I'll bring out our own dinners in a few minutes, but please start and enjoy your fish while it's hot," Ms. Karen said.

Everyone tasted their own order of perch and then took a piece of fish from the plate to their left and then from the right. Once those were tasted, pieces were offered from the remaining two plates.

Ms. Karen watched the family through a window in the kitchen door while waiting for their own dinners to be prepared. She admired the orderly way in which everyone was able to taste everyone else's fish, thanks to Grandpa Mike's clever direction. *He must be a schoolteacher*, she thought.

Ms. Karen came out carrying two more fish dinners and placed them at the two empty places at the table.

"We're delighted you're able to join us," the twins' Mom said. "We'd love to hear more about the area from people who actually live here."

"We're third generation innkeepers," Mr. Tom said. "Karen's grandfather built this inn in the early 1900s, then her mother and father ran it, and now we run it. Most of the wood you see here is the original timber; we've tried to preserve everything we can, while adding modern touches such as the free Wi-Fi. People can't seem to get enough of the Internet, even on vacation."

"How true," said Grandpa Mike. "That's why we've decided to make our time screen-free when we're out enjoying nature. There are some recent studies that show our brains are more active when we use our own senses instead of relying on GPS and other apps all the time."

The family looked around, admiring the craftsmanship surrounding them.

Then Mallory said, "Ms. Karen, I really like the fish with the stuff sprinkled on top. What's it called?"

Fish had not been on Mallory's list of favorite foods...until now.

"It's parmesan cheese mixed together with herbs and a special kind of crumb, like bread crumbs," Ms. Karen answered. "It's our cook's special recipe. We'll be sure to tell Fred that you like it. I'll ask him to write out his recipe and give it to you before you leave the island."

"That's neat, Ms. Karen. Thank you, ma'am."

Mr. Tom said, "I haven't forgotten your question, Mike: *How long and wide is Washington Island?* I'll draw a rough map on the back of this placemat." After he drew the outline of the island, Mr. Tom drew two lines, one east-to-west and the other north-to-south. "The island is about five miles long and about 5 miles wide." Then he wrote the number "5" next to each of the two lines.

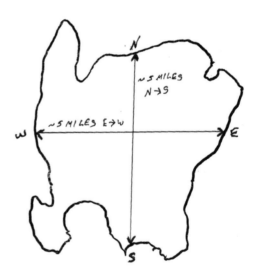

Grandpa Mike asked, "Mel, Mal, on the ferry your Dad told your Mom that the shoreline wasn't far from wherever you might be on this island. Can you figure out what's the farthest you can be from water if you could walk in a straight line from where you are until you reach water?"

Brother and sister looked up and down, left and right. Two sets of eyes settled where the two lines met. "Two and a half miles," they both said at once.

"Excellent," Ms. Karen said. "Please tell us how you figured it out so quickly."

Melody quickly pointed to the west-east line. "See? It's the same distance from side to side. So is the up-down line, except for those places poking out on the west side."

Mallory then put his finger on the point where the two lines joined. "That's the middle of both lines, so the middle of the line is the farthest you can be from the water. Middle means halfway, and half of five is two and a half."

The twins' Mom and Dad, Grandpa Mike, Ms. Karen and Mr. Tom smiled and clapped their hands at the answer. Since everyone had finished their dinner, Ms. Karen announced dessert. "I hope you've all saved room for our famous cherry desserts. Let me bring out samples. And guess who gets to pick first!"

Everyone knew who got to pick first. It was a tie.

# Chapter 2
# Washington Island: First Explorations

## Schoolhouse Beach

After breakfast at the inn's restaurant, everyone piled into the car for the short drive to Schoolhouse Beach.

"I want to build a rock skyscraper," Mallory announced.

"I want to build the prettiest rock castle ever built," Melody said.

Rufus barked his approval. Who knows? Maybe he felt the twins' excitement and was ready to enjoy his own kind of fun. Knocking down whatever Mallory or Melody built was what he did best.

## New Friends

A group of boys ran up as soon as they saw Rufus jump down from the car. "Can we play with your dog? What's his name?" they asked.

"Certainly, boys," said the twins' Mom. "Our dog's name is Rufus. This is Mal, short for Mallory and this is Mel, short for Melody. Are you from around here? What are your names?"

"We're cousins and we come up every year. We live close to Chicago," said the oldest boy. "I'm Bobby, and this is Fred, Tim and Steve. My parents and my aunt and uncle are over there," he said, pointing to a picnic table next to some tall trees.

The adults waved hello when they saw Bobby pointing in their direction.

"It's nice to meet you," Grandpa Mike said. "This is our first visit here."

"We're also from the Chicago area," said the twins' Mom. "Maybe you children can get acquainted while we older folks do the same."

"Be sure not to let Rufus run wild," the twins' Dad warned.

"We won't", both Melody and Mallory said together.

"How old are you?" Mallory asked. "I'm seven, and so is my sister."

"I'm twelve," Bobby said.

"I'm eleven," Fred said.

"So am I," said Steve.

"I'm ten," said Tim.

"Can you show us how to skip stones?" Melody asked. "Mr. Tom, where we're staying, said skipping stones is fun."

"Sure," Tim said.

"Let's go down by the water," Bobby said as he ran ahead of the others.

Bobby had already picked up a stone and threw it sideways toward the water. The stone skipped twice as it hit the water before finally sinking.

"Aw, that wasn't so good," Bobby said. "Let me try again."

Rufus was already in the water before Bobby threw the second stone. Rufus tried to catch the stone, but missed each time as it skipped three times over the surface.

"That wasn't very good either," Bobby complained.

"Rufus, come here," Melody called. "You're messing up Bobby's throws. Besides, you could get bopped in the head!"

"Ha, ha. I don't think so," Steve said. "Here, watch this."

Steve picked up a stone and threw it really hard. It skipped four times before sinking.

"Okay, you guys try now," Bobby said. He wasn't exactly happy that his younger cousin did better than he did.

Mallory picked up a stone and threw it as hard as he could. Kerplunck. It sank without skipping at all. Rufus watched as the stone disappeared and ran back to the stone beach. He shook himself off, getting everyone wet. Nobody minded, though. It was a warm day and the cool water felt good.

Melody picked up a stone and massaged it with both her hands. "That's for good luck," she said. She tried to imitate the way Bobby and Steve and thrown their stones. It skipped once before sinking. "See? That's how it's done," she said to her brother.

"Let's line up and throw stones together," Tim suggested. The four cousins and the twins lined up along the shore, while Rufus ran into the water right in front of them.

"Rufus, come here," Bobby said.

Rufus was surprised to hear a new voice calling his name, so he obediently hopped out of the water. He was even more surprised when Bobby stepped on his leash instead of handing him a treat.

"Ready, set, go!" Bobby shouted as six stones flew through the air. Two stones clunked together and dropped together into the water. One stone skipped four times, beating the others. The problem was that no one was paying attention to which stone was whose, so everyone was declared the winner.

Fred then asked the twins, "Want to see something really neat?"

"Sure," they both said.

"We'll show you," Fred said. "Timmy, would you go and tell Mom and Dad where we're going?"

Timmy didn't need to guess what Fred meant, so he ran off to the picnic bench, told his parents where they were going, and then ran to catch up with the others.

Melody stood and stared. "Oh, wow! A stone city! That's neat! How did all these stone castles get here?"

"I don't know," Bobby said. "They were here as long as I can remember. Let's ask when we get back."

Melody and Mallory took turns trying out different pieces of rock furniture while their new friends pointed to their own favorites.

"Can we play musical chairs?" Mallory asked.

"As long as you don't mind bruised shins or even worse if you fall down," Bobby said. He was speaking from experience. His cousins laughed as they remembered their own encounters with rock chairs. They were NOT soft like regular furniture *rocking chairs.*

"Let's go back and ask if our folks know who made these," Steve suggested.

"Good idea," Bobby said. "Then maybe we could go to Little Lake. It's close by."

"Mom, do you know who made those castles and other neat stuff out of the rocks?" Bobby asked, as all six children ran up to the picnic table where the adults were talking and getting to know each other.

"No, I can't say that I do. Do any of you know?" she asked the others.

"We saw the neatest stuff make out of rocks," Mallory said to his parents and Grandpa Mike.

"That sounds interesting. I'd like a look," said Grandpa Mike, as everyone else shook their heads.

As the adults got up from the picnic bench, the twins' Dad asked, "Shouldn't we be taking our things with us?"

Bobby's Mom answered, "Don't worry. They'll still be here when we get back."

Lots of interesting rock sculptures were everywhere, some small and simple, others quite large and finely detailed. "These smooth rocks bring out many people's creativity. They seem to enjoy building and arranging the rocks, and then leaving their artistry for others to enjoy," said Tim's Mom.

"Dad, could we go to Little Lake? I think Mal and Mel would like to see it, and it's not on too many maps," said Bobby.

"That's a good idea, Bobby," his Dad said. "It's got a nice little trail, although it might be a bit difficult for the youngsters."

"Mel and I can do it if Tim can," Mallory said.

Tim nodded. He felt proud to be able to let his new friends know that he was able to hike the trail.

"If it's okay with everyone, I can be the lead driver and the rest of you can follow in your own cars," said Bobby's Dad.


**Little Lake**

"Little Lake is the only inland lake on Washington Island," Bobby's Dad told the Maloney family. "It's not very deep, only about six feet, I understand."

"How does it get its water?" Melody asked.

"From rain, silly," Mallory answered.

"Rain is just a small part of where Little Lake's water comes from, Mal," Tim's Mom said. "My family has lived here for a long time. My Dad worked for the Nature Preserve here, so I learned a whole lot about Little Lake. Actually, this lake gets its water mostly from underground springs."

"We learned about underground springs when we were at Yellowstone," Mallory said. "Those were mostly hot, though, with the water coming up from way underneath the earth."

"And Uncle Frank drew us a picture of groundwater. That was at his and Aunt Martha's farm," Melody said.

"Sorry for my children's chattering, Susan," said their Mom. "Mal, Mel: remember, you learn better with your ears than your mouth."

"That's quite all right, Agnes," Tim's Mom said with a smile. "I'm surprised and also happy that you know so much, Mel and Mal. The six of you youngsters will have a lot to learn from each other."

As they walked along the hiking trail, the younger and older "old timers" eagerly pointed out different sights to their new friends and explained some of the history behind Little Lake and its present condition:

- Little Lake was formed thousands of years ago when glaciers receded, creating a ridge to contain the lake;
- Native Americans lived in this area for more than 3,000 years, based on human-made items found here; Little Lake also has a Native American burial ground;
- The trees around Little Lake are mostly white cedar and hemlock; there is a floating bog mat that supports plants preferring marshy conditions;
- Parts of Little Lake and its shoreline are now protected by Door County's Land Trust; this means that it will remain undeveloped to preserve its natural environment;
- Interesting birds, such as bald eagles and white pelicans, can be seen feeding and even nesting during some seasons;
- Little Lake is only about three feet above the level of Lake Michigan; although Green Bay is on its western edge, you can see Lake Michigan from a certain point along the trail.

"Mom, I'm getting hungry," Bobby said when they had finished walking the trail.

Mallory's Mom understood. She looked at her own son, who was nodding his head. "Growing boys need to be fed, don't they?" she asked.

"Girls, too," Melody said.

"Shall we all enjoy lunch together?" Grandpa Mike asked.

Everyone thought it was a great idea. Mallory wanted to ask the boys about Death's Door, and the four boys wanted to hear more about groundwater and other things that Mallory and Melody talked about.

"It's getting hot," said Tim. "Can we go swimming at Sand Dunes Beach after lunch?"

"Wow, I'd really like that," Mallory said. "I'm getting hot, too."

"Where are you staying?" asked Tim's Dad. "Sand Dunes is on the south end of the island, on the other side of Detroit Harbor where the ferry comes in from the mainland."

"We're staying at an inn about two miles from the Welcome Center," said Mallory's Mom.

"Our place is nearby; it has a nice little café where we can get lunch before picking up our swimming gear," said Tim's Dad.

"Or we can pick up some food for a picnic lunch at the beach," Tim's Mom suggested.

"That sounds like fun," Grandpa Mike said.

### Death's Door Lessons at Sand Dunes Beach

"The water level looks high enough for good swimming," Bobby's Mom said as the families arrived at the beach. "Sometimes it's so low that you have to walk too far out to get to the water. The nice thing about this beach is that it has a nice, gentle slope so even little ones won't get in over their heads."

"We're not little," Melody protested.

"No, you're not," Bobby's Mom agreed. "I was talking about youngsters half your age," she said with a wink to Melody's Mom.

The families gathered around a large picnic blanket spread with different kinds of sandwiches, carrot and celery sticks, a leafy salad, and assorted fruit. Bobby's Dad kept everyone's water cup filled from two large jugs.

After lunch, while they were waiting to let their food settle before going swimming, Mallory asked the four cousins, "What's Death's Door like? We kind of missed it when we came over on the ferry. It sounds spooky, in a neat way."

 Death's Door appeared in both Mallory's and Melody's dreams the night before (although they appeared in different shapes). Mallory dreamt of a skull and crossbones floating above choppy waters. Melody dreamt of a ghost in a raggedy white sheet. Neither of their dreams was scary, though, just interesting.

"Lots of shipwrecks are down there, in the water on the bottom," Fred said.

"With lots of treasure," said Steve.

"What kind of treasure?" Melody asked.

"Oh, gold and silver, maybe jewelry. Stuff that people brought with them for trading," Bobby said.

Their parents listened to what the boys were saying. Then Melody's Mom said, "There were sailing boats from France, if I recall what the ferry captain said. Sometimes the waters were so rough, their boats tipped over and sunk. I read somewhere that many Native American canoes were sunk as well."

"What caused them to sink?" Tim asked.

"Ah," said Grandpa Mike. "We can use the work of a physicist and mathematician named Bernoulli. He lived more than two hundred years ago. Who can tell me where Death's Door is?"

"It's between the mainland and Washington Island," Steve said.

"Good. Now, who can tell me what two bodies of water are on either side of the mainland and Washington Island?" Grandpa Mike asked.

Fred raised his hand and said, "Green Bay is on one side and Lake Michigan is on the other."

"You boys know the geography of this place very well," Grandpa Mike said, as Melody and Mallory wondered what Grandpa Mike would ask next.

"Since Green Bay and Lake Michigan are two separate bodies of water, they must be connected somewhere, right?"

Six young voices said at almost the same time, "Death's Door!"

"Excellent!" said Grandpa Mike as he looked at the children. "What makes Death's Door so different from Green Bay or Lake Michigan?"

"It's lots smaller," Steve said.

"It's narrow," Bobby said.

"You're both right," Grandpa Mike said. "This is where Mr. Bernoulli's investigations come in. He noticed that liquids and gases behave almost the same when they encounter narrow openings. What liquid do we have at Death's Door?"

The same six voices shouted, "Water!"

"And what gas do we have at Death's Door?"

Silence.

"Um, air is a gas, right?" Bobby asked.

"And wind," Fred said.

"Wind and water. Hmm, that sounds like a storm, doesn't it?" Grandpa Mike asked. Everyone nodded.

"What do you suppose happens when you try to squeeze a lot of wind and water through a narrow passage, such as Death's Door?"

Bobby smiled. He knew the answer. He put his finger to his lips as if to say "shh" when he saw that Steve and Fred knew the answer, too.

Tim's eyes got big when he realized he knew the answer to the question he asked earlier. "It goes faster. And the faster wind and rain caused the boats to sink," he said proudly.

Melody and Mallory clapped their hands. So did Tim's cousins, and then their parents joined in.

Grandpa Mike felt pretty proud of all the children, too. "Would you like to do a simple experiment that shows how it works?"

Everyone nodded, and not only the youngsters.

"Hold your hand up in front of your face, as though you wanted to cover a yawn. Now, open your mouth and blow into your palm," Grandpa Mike said, as he demonstrated.

"Okay, now pretend you're blowing out candles on your birthday cake. Don't blow any harder than you did the first time. Do you feel the difference in the speed of your breath on your palm?" he asked.

"That's neat, Grandpa," Melody said. "I could really feel the difference."

"My breath came out lots faster," said Tim.

"Now you can understand how dangerous narrow waters can be," Tim's Mom said.

"It's the same with wind hitting tall buildings in a city," Bobby said. "It's a lot windier in the narrow streets than out in the open."


## Swimming at Sand Dunes Beach

After Grandpa Mike's physics lesson, everyone was ready to go in the water. Waiting a half hour after eating before going into the water was never so much fun.

"Last one in is a rotten egg," Fred shouted.

"Let's pretend that we're at Death's Door and I'm a big storm," Mallory yelled. He belly-flopped into the shallow water.

SPLASH!!

Then, "OUCH!"

Mallory learned the hard way how hard water could be when you hit it flat. Unfortunately, only Rufus was nearby, so Mallory was the only human who got wet.

His Mom saw what happened and waded up to him. "That was quite a belly-flop, Mal. Your belly and chest should stop hurting in a minute. Next time you want to make waves, do a cannonball. It makes a bigger splash and it doesn't hurt."

"What's a cannonball, Mom?"

"I'll show you. It's too shallow here. We need to go into deeper water. I don't want to hit the bottom with my bottom."

Mallory and his Mom waded out until the water was up to her waist, which means the water was up to Mallory's shoulders.

"Watch this," she said.

Mallory watched as his Mom couched down, low into the water. Then she jumped up as high as she could. She used both hands to pull her knees to her chin when she was still in the air and landed in the water. She made a much bigger splash than Mallory's.

"There," she said. "It's much easier if you have a diving platform, but you get the idea."

Mallory's Dad and Grandpa waded over, followed by Melody.

"That was some jump, Mom," Melody said.

"It sure was," Mallory's Dad and Grandpa said at the same time.

No one else seemed to have noticed Mallory's little splash or his Mom's much bigger one.

After an hour or so of tossing a big beach ball to one another in the water, everyone gathered on the sand to see who could build the best sand castle or other creation before Rufus would knock them down.

"Were you planning on visiting Rock Island while you're here, Agnes?" Bobby's Mom asked. "I think your children would enjoy a day or two there. I'm sorry that we aren't able to join you because we have several family visits to make over the next several days. I'm sure your twins will have great fun exploring on their own. Here, let me write down my cell phone number so we can stay in touch."

"Thank you, Kris. Here's my cell phone number. As a matter of fact, we were thinking that Rock Island would be an interesting place for us to explore," said the twins' Mom. "We'll probably take the ferry over tomorrow. I understand that no vehicles are allowed, not even bicycles. Is that right?"

Bobby's Mom answered, "Yes, the county wants to keep the island in its natural state as much as possible. That's why there are only campgrounds on Rock Island, no inns or food service. I'd advise you to pack lunches for yourselves, but don't bring anything heavy. What you bring in you also need to carry with you back out."

Bobby's Dad said, "There are lots of interesting hiking trails for you to choose from, especially near where the ferry drops you off. You might want to save going to the Pottawatomie Lighthouse for another day if you decide to explore those close-in trails first. You get to the lighthouse by taking the trail on the west side of the island. There are free guided tours of the lighthouse every day."

"I want to go to the lighthouse first," said Melody.

"Me, too," Mallory said.

"First things first," their Dad said. "We need to thank these nice people for sharing their vacation time with us."

Mallory and Melody shook hands with the four cousins and then with the adults. They told everyone other how much fun they had and that they hoped to see each other again. The cousins even shook hands (paws?) with Rufus.

As they headed back to their cars, Grandpa Mike suggested, "Let's wipe off 'sandy-paws' Rufus,' so he won't scatter sand all over the car."

"Good idea. Let's also brush the sand off of all of *us* before *we* scatter sand all over the car," their Mom added.

# Chapter 3
## An Even Smaller Island

"So, you're planning on visiting Rock Island tomorrow," Mr. Tom said to the twins as he led the family to a table for dinner. Their Mom had asked if he had any advice on the best way to enjoy the island.

"We're going to see the lighthouse," Melody said.

"That's a fine idea," Mr. Tom said. "Just be sure you don't miss the last ferry of the day coming back here to Washington Island – unless, that is, you want to camp out under the stars with no food in your bellies. I'll bring you a schedule of the Karfi ferry. You might want to plan on going out there for a second day, though. There's a lot to see and do on Rock Island."

Mallory didn't think he'd mind sleeping out in the open. No dinner, though? Unthinkable! "May I look at the ferry schedule?" he asked when Mr. Tom brought the sheet to his Dad. Even though he didn't wear a watch, Mallory wanted to make sure he knew when the last ferry left Rock Island.

"Oh, one more thing," Mr. Tom said. "The whole island is a completely primitive place. There are no shops, so be sure to bring enough food and water containers. You can fill your water bottles from hand pumps that are in a couple of places."

"What if we order lunch and snacks now and pick them up after we finish breakfast tomorrow?" Grandpa Mike asked.

"I'll be happy to have everything ready for you. Just let me know what time you'd like them to be ready," Mr. Tom said.

"Let's take the first ferry out," said Melody.

"And the last ferry back," said Mallory.

\*\*\*\*\*\*\*\*\*\*\*\*\*\*\*\*\*\*\*\*\*\*\*\*\*\*\*

## On the Way to the Lighthouse

Everyone was up early the next morning. Seeing a lighthouse up close would be a first experience for everyone. Even better, Mr. Tom had told them that there would be guides to take them through the lighthouse itself, which had been transformed into a museum.

"Backpack check!" said the twins' Dad, before they left the inn.

Even Rufus had his own doggie backpack. His backpack was like a harness with two bags to hang on either side. Rufus wasn't exactly keen on the idea, but he got to carry his own collapsible water bowl, along with people's items such as plastic forks and washcloths.

Melody patted him on the head. "You look like a little pony, Rufus. At least you don't need to carry your own food. You just ate breakfast and can wait until we get back for dinner," she told him.

Rufus just shook his body all over and snorted.

"Be sure to keep your pooch on his leash," one of the passengers told her Dad.

"And keep it short," another passenger advised. "Dogs like catching the spray of the waves as much as they like sticking their heads out of car windows. Before you know it, a big wave comes by and SLURP! They're gone."

Once Rufus got the idea that they were going on a new adventure, he didn't pay any more attention to his doggie backpack.

The ferry ride was calm, and the passengers found themselves on Rock Island in less than ten minutes. All the Maloneys agreed that they should visit the Pottawatomi Lighthouse first, especially since the last ferry of the day back to Washington Island was scheduled to leave just eight hours later.

Mallory pointed to the Rock Island map. "Let's take the west part of this Thordarson Loop Trail to the lighthouse. Then we can take the longer way back if we have time."

"Are you afraid of missing the last ferry back and missing dinner?" Mallory's Mom asked, guessing what was on her son's mind.

Mallory just grinned.

"Let's fill up our water bottles here," Grandpa Mike said, as they passed hand pumps.

"It's almost like the one we used on the farm, Grandpa," Melody said.

"So that means you don't need any help working them, right?" her Dad asked.

"Nope," Mallory answered, as he started pumping the handle up and down until cold water began flowing.

First, Melody filled up her own bottle, then her Mom filled hers.

Pump, pump, pump.

Mallory kept pumping. His Dad then filled his own bottle and then another bottle to fill Rufus's collapsible water bowl.

Pump, pump, pump. Slower and slower.

Grandpa Mike said, "Here, Mal, you can fill our two water bottles while I pump."

*Grandpa really knows everything*, Mallory thought. *He didn't want anybody else to know that my arms were getting tired.*

"Thanks, Grandpa!"

Supplied with enough drinking water, the Maloneys enjoyed a leisurely walk along the trail to the lighthouse. Green Bay looked so beautiful and peaceful...and green...in the morning light.

At last they came to the lighthouse.

The twins were excited to see both Green Bay and Lake Michigan as they stood at the northern edge of Rock Island. The steep bluff reminded them of the Mississippi Palisades.

"Just think about it," said their Mom. "You could stand in the same place and see the sun rising over the water by looking to the right in the morning, and see the sun sinking into the water to the left in the evening."

"Wow. Neat! Is there any other place in the whole world like this?" Mallory was fascinated with the thought of standing in one place and being able to see the sun rise and set over the same body of water.

Melody was disappointed. She expected to see a very tall tower with a light on top, like the kind you see in pictures. She didn't expect to see a stone building that looked like a house that also happened to have a small tower to hold a light inside its clear top.

"That's just right for our picnic lunch," Mallory said, pointing to a grassy area to one side of the lighthouse entrance. "We can eat first, and then take the tour, right?"

Lunch didn't take long to eat. As they walked up to the lighthouse door, a tour guide greeted them.

## Introduction to Rock Island's Historic Lighthouse

"Welcome to Pottawatomie Lighthouse," she said. "My name is Louise, and I'm a volunteer docent here at the lighthouse."

"What's a docent?" Melody asked.

"Some people say *docent* is just a fancy word for *tour guide*," she said, smiling. This was a question that she heard at least once a day. "I think it comes from the Latin word for *teach*, but I'm not sure."

"We do a lot more than give tours," Louise added. "We also clean the lighthouse and outhouse, and keep the grounds clean. We pump water by hand for cooking and washing up. Everything is like it was in the 1800s, except for the refrigerator and stove that are run with propane, some battery-powered camping lights, and a brand new solar pump to help pump the water from a new well."

"That sounds like a lot of hard work," Melody's Dad said.

"It is, but it's only for a week at a time, so it's not so bad."

"Hello, I'm Brian," said a tall young man who had entered from a side door. "Louise is an expert on the lighthouse and its light. She'll explain how lenses work to magnify a light so the light can be seen for many miles over the water."

Louise looked outside. There were no other visitors coming up the path, so the Maloneys enjoyed their own private tour of the lighthouse. Louise led them to the living areas of the lighthouse. Grandpa Mike tied Rufus's leash to a nearby post.

"Did you know you're in Wisconsin's oldest lighthouse?" Louise explained that the original lighthouse was replaced in the 1800s and had been restored more recently. As they climbed the narrow steps up to the tower, Louise explained the importance of lighthouses, how they shine great distances to keep ships from running into land when it's dark. "Not everyone uses GPS devices, so lighthouses still serve a very valuable function. Before lighthouses, people built big bonfires on top of hills so ship captains would know where land was."

Grandpa Mike added, "Airplane pilots also use lighthouses as visual markers. GPS technology is great, but your eyes are one of the best tools you can use."

"This is the lantern room. The actual, working light has been moved to a steel tower in front of the lighthouse," Louise explained. "That's because trees have a habit of growing, and the Wisconsin Department of Natural Resources wouldn't allow them to be cut down. The light is automated, so there's no longer a need for a permanent keeper. What you see here is a plastic replica of the actual beacon."

"How can a light shine so far away?" Mallory asked.

"I'm so glad you asked that, young man," Louise said. "This is a replica of the lighthouse's Fresnel lens. It's pronounced *fre-nel*, even though it's spelled with an *s* in the middle of the word. It's named after the French physicist who invented this type of lens."

Louise continued, "Let me give you a simple explanation of a Fresnel lens. If you look closely, you can see that there are lots of rings, similar to the rings of a tree. The angle of each ring is different, and they all work to focus the light toward the center. This focus to the center works to magnify the intensity of the original light so that it shines a much greater distance than without the lens."

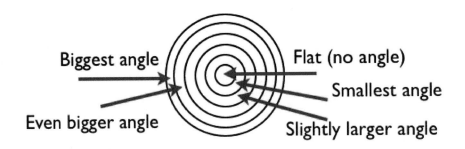

"Um, okay," Mallory said. He wasn't quite sure of everything Louise said. Neither was Melody.

Their Dad said, "Fresnel lenses have been used in car headlights to focus the beam; I understand that some solar cell makers use Fresnel lenses for concentrating sunlight."

"You can use them to start a fire and even cook food," their Mom said. Mallory grinned. Cooking seemed like a good idea to him.

"By the way," Louise said. You should be aware that the spelling of *Pottawatomie* Lighthouse is different from the spelling of the *Potawatomi* Native Americans. Our lighthouse has two *t*'s and the Native American group is spelled with only one *t*."

After the guided tour of the Pottawatomie Lighthouse, the family planned to follow the Thordarson Trail around the east part of the island and then back west and north to the ferry.

Just beyond the lighthouse, wooden steps led down from the top of the cliff to the beach. Melody and Mallory ran with Rufus to the edge of the stairway.

"That looks like a great place to explore, Mel," Mallory said, pointing downward.

"That's awfully steep," said his Dad as the rest of the family gathered around. "Rufus would never make it down there. In this case, we also need to remember, *whoever climbs down, must also climb back up.*"

"Gravity will get you every time," Grandpa Mike said, smiling.

Mallory's Mom looked at her watch. "*Time and tide wait for no man,* the saying goes. The clock to the last ferry is ticking away, and those steps look a bit dangerous. What would happen if anyone slips and falls? We'd have a hard time getting help."

"Aw," Mallory said.

Mallory's Mom knew how to distract her son. "If we have time, we can check out the Blueberry Trail."

"I like blueberries. Can we pick some?"

"It might be a bit late in the season for wild blueberries, Mal," Grandpa Mike said.

"And other people might have already picked all of them," said Melody. *Food is okay,* Melody thought. *I'm not a 'food hound' like my brother and Rufus, though.*

"I'd also like to check out the Thordarson Boathouse," said the twins' Dad. "The brochure makes it sound pretty impressive."

"Well, let's see how far we get before the last ferry leaves for Washington Island," their Mom said as they continued walking.

Just a short distance away, a scenic overlook came into view.

"*Water, water everywhere, and all the boards did shrink,*" their Dad said, quoting a famous poem about a long sea voyage.

"'*Water, water everywhere, and not a drop to drink,*'" the twins' Mom continued.

"'*The Rime of the Ancient Mariner,*' by Samuel Taylor Coleridge," said Grandpa Mike as he walked along the overlook, admiring the beautiful view. He waved at the wide stretch of water sparkling in the sun. "Fortunately, this is drinkable lake water, not salty sea water…".

# Chapter 4
## An Unplanned Delay

Suddenly, Grandpa Mike stumbled.

Fortunately, he was able to regain his balance before falling

*Unfortunately*, Grandpa Mike could only stand on one leg. His other leg couldn't support even a little bit of weight without causing intense pain.

"What happened, Grandpa?" Melody asked, as everyone ran up to help him sit down on a nearby bench.

"It feels like I sprained my ankle when I stepped the wrong way on that stone," Grandpa Mike said as he pointed to a fairly large stone that was sticking up out of the ground. "I should know better than to talk and walk and think, all at the same time."

The twins' Mom said, "Mal, Mel, go run back to the lighthouse and ask for help. There should be at least one guide at the entrance. Tell them your grandpa probably sprained an ankle. They should have a first-aid kit. We may need someone to come and help Grandpa back to the lighthouse."

"And take Rufus with you," their Dad added.

Rufus seemed to know where to run; he reached the lighthouse and stood barking at the front door when Melody and Mallory caught up with him.

"What seems to be the trouble, kids?" Louise asked.

"Grandpa tripped over a rock and hurt his ankle," Melody said.

"Wait right here. I'll get Brian and we'll both come right away. Where did it happen?"

Mallory answered, "He's sitting on a bench. It's not too far from here. We ran all the way. Mom and Dad are with him."

In a few minutes, Louise and Brian came out of the lighthouse. Brian carried wet towels wrapped in plastic to help keep swelling down; Louise carried a walking stick.

It took only a few more minutes for all of them to reach the park bench. Grandpa Mike had already taken off his shoe and sock. Brian knelt down and looked at Grandpa Mike's ankle. He saw that it had already begun to swell a little.

"Do you think it's broken, sir?" he asked.

"I don't think so," Grandpa Mike answered. "I've broken bones in the past, and I didn't feel the electric-shock kind of pain that a bone break causes. I'm pretty sure it's just a sprain."

"Unfortunately, we don't have a hospital here on Rock Island, so we can't be sure," Louise said. "I'll get in touch with the ranger who lives near the ferry dock and let him know what happened. We have emergency helicopter service..."

Grandpa Mike protested. "No, no. This is not an emergency. I should be fine soon. The cool towels you brought should do the trick."

"Our guidelines are that we have to call the ranger whenever there's any kind of accident, sir," Brian explained.

"I understand. I won't stand in the way of your policies," Grandpa Mike said. "A bit of rest, some aspirin, and I'm sure I'll be fine."

Brian and Louise wrapped the plastic bags with the wet towels around Grandpa Mike's ankle.

"Ah, that feels good," Grandpa Mike said. "I should be good as new in no time."

After about five minutes of additional rest, Grandpa Mike felt ready to test his ankle. "Let's see how far I can walk," he said.

The twins' Dad and Brian helped Grandpa Mike to stand up. Their Mom picked up his shoe and sock. Once Grandpa Mike was standing, Louise gave him the walking stick.

"Take as long as you need to, sir," Brian said.

"Thank you, Louise and Brian. I really appreciate your help," Grandpa Mike said. "By the way, we're not wearing name tags as you are, so let me re-introduce you to our family: Melody, Mallory, Agnes, Mort and Rufus. And I'm Mike."

Louise called the ranger while the two men, one on each side, helped Grandpa Mike walk slowly back to the lighthouse. Mallory and Melody walked behind them, holding Rufus by his leash.

Mallory pointed to his Mom's watch and whispered, "What time is it, Mom?"

"Shh," his Mom said. "First things first." She knew Mallory was worried about missing the last ferry back to Washington Island (and worse, missing dinner). She was worried as much as Mallory was, but she felt that somehow things would work out.

## Bad News/Good News

"I've got some bad news and some good news," Louise announced when she finished speaking with the ranger.

"Let's have the bad news first," said Grandpa Mike. "I'm the one who caused all this trouble, so I hope we can get that part over with."

"Okay, the bad news is that all the larger boats with enough capacity to pick up five passengers and a dog are unavailable to come out this way. With Grandpa Maloney unable to walk all the way back to the ferry landing this afternoon, there is no chance of catching the last ferry back to Washington Island."

"What's the good news?" Melody asked.

"The warden has requested that two of the docents on duty take a small boat back to Washington Island later this afternoon. You will be able to stay the night in the lighthouse and use their beds. Melody, you and your Mom can have one bed, your Dad and brother the other bed. We'll unfold our spare cot. This way your Grandpa can sleep alone, so his foot can be undisturbed during the night. Brian and I will be here to take care of all of you. Let me know where you're staying on Washington Island and I'll call to let them know that you won't be returning today."

"That's very kind of you, Louise, thank you," said her Mom.

"What's for supper?" Mallory asked.

"Mallory Maloney," his Mom scolded. "What a thing to ask. You should be concerned about your Grandpa, not about your belly."

"*Gotta eat,*" Mallory thought, but dared not say aloud.

"Don't worry about food, young man," Brian said. "After all, we have to eat, too. We've got plenty of food to keep you happy."

"*Don't count on it,*" thought Mallory's Mom, smiling to herself.

Louise said, "We don't get overnight visitors that often. Suppose we have a sing-along after we eat. Once it gets really dark outside, we can enjoy the beautiful night sky and watch for the International Space Station as it flies over us. We've got a schedule that shows when we should be able to see it. We'll have to watch carefully, though. It's there and then it's gone in less than a minute."

Melody was confused. "That's where astronauts stay, high up in the sky, right? How can you see the space station?" she asked. "It's not a star."

"That's right, Melody, it's not a star," said Brian. "The space station is a lot closer to us than the stars are, so it will look about the same as a star. It's bright enough to see because it reflects light from the sun. You can tell it's the space station because you can see it moving, while the stars look like they're standing still."

"How are you doing, Dad?" Melody's Mom asked. She didn't want her father-in-law to think no one cared about his sprained ankle.

"Could be better, my dear. I'm embarrassed to be causing all this trouble."

Mallory quickly said, "It's okay, Grandpa. It'll be fun to stay and look at all the stars and watch the space station go by."

His Mom poked him, pointing to Grandpa Mike's foot.

"I'll bet your foot will be all better tomorrow," he added.

"Can we watch the sun go down on one side of the water and come up tomorrow on the other side?" Melody asked.

"That's a great idea, Melody. I'd guess your Mom and Dad will want you to go to bed right after watching the space station go by."

"Thank you, Louise," said Melody's Mom. "I suspect you'll make a great Mom someday."

Supper was a simple meal of salmon caught that morning, baked beans, carrot sticks, and boiled potatoes. Rufus got generous helpings of the salmon and potatoes along with some dry dog food that Louise found in the kitchen pantry.

Everyone helped clean up the kitchen when they had finished eating. Even Grandpa Mike helped by drying the dishes. He sat on a chair with his leg propped up on a stool. Louise got him a cool, wet towel to wrap around his ankle while he dried each dish as it was passed to him.

Once the kitchen was completely spotless, everyone went outdoors to enjoy a desert of oatmeal cookies and Door County cherries. Rufus got some chewy doggie treats.

*I'll have to make sure Rufus's share gets replaced with fresh pet food,* thought Grandpa Mike.

Louise and Brian each brought out a guitar.

"It's funny. We didn't know each other before volunteering to be docents here, and it turns out we both play guitar and just happened to bring them with us," Louise said.

"It's also funny that we both like the old sing-along songs that just about everyone knows," Brian said.

After *Row, Row, Row Your Boat* and a dozen other simple songs, Louise suggested that she and Brian sing a song that she wrote while putting the dishes away. They sang it to the tune *On Top of Old Smoky:*

> On top of Rock Island,
> A lighthouse does stand
> And Grandpa Maloney
> So hard he did land.
>
> The rock that he slipped on
> His ankle did twist
> A sprained ankle is worse than
> A sprain of the wrist.
>
> Let's hope that it heals well
> And be good as new
> So, Grandpa Maloney,
> Don't slip on the dew.

"Wonderful!" Grandpa Mike declared. "I feel that my ankle is getting better already."

His ankle wasn't getting better, but Grandpa Mike didn't want to spoil his family's unplanned adventure this evening.

"Your leg is still all puffy, Grandpa," Melody said. "May I feel it?"

"Alright, if you want to," Grandpa Mike said as he removed the wet towel draped around his ankle.

Melody knelt down and cupped her hands around both sides of Grandpa Mike's swollen ankle.

Melody could feel heat flowing from Grandpa Mike's ankle into her cool hands. Grandpa Mike could feel the coolness from Melody's hands flowing into his ankle. To both Grandpa Mike and Melody, it felt like two different layers of clouds, one warm and one cool, moving in opposite directions.

Even after Melody removed her hands, she still felt the warmth of Grandpa Mike's swollen ankle, and Grandpa Mike still felt the coolness of his granddaughter's hands on his ankle.

Grandpa Mike told Louise he wanted to speak to the ranger and asked if he could phone him.

"Of course. Here is my phone. The number is on speed dial."

Grandpa Mike stepped inside to make the call. He described his idea to the ranger.

"Thank you for your thoughtfulness and generosity, Mr. Maloney. I'll arrange all the details," the ranger said.

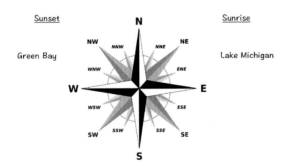

Grandpa Mike finished his call about ten minutes before sunset. Brian brought out a compass so they would know the exact point where the sun set and be able to compare it with the point on the compass where the sun would rise the next day. He also brought out several paper copies of a commonly-used compass diagram with labels added for the locations of sunrise, sunset, Green Bay, and Lake Michigan.

"Let's stand facing *north* so we can identify on the compass where the sun sets," Brian said. He had also brought out a red marking pen so that Melody and Mallory could mark the sheets and take them home.

Melody and Mallory took turns using the compass to find the direction of the setting sun. They made small marks on one of the sheets, and when they agreed on the correct compass reading, Melody put a star where the sun had set. They figured out (more or less) where the sun would arise the next morning, but they realized that they had no choice but to wait until the sun actually rose over Lake Michigan in the morning.

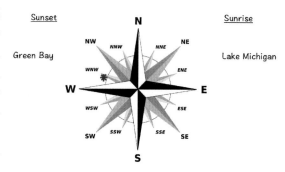

As the sky darkened, stars began to appear, one-by-one at first, and then an explosion of tiny and not-so-tiny points of light.

Louise checked the schedule of when the space station should appear. "It should be visible in another ten minutes," she announced.

Mallory was the first to see the space station. "Is that it? It's moving and the other stars aren't!"

"That's right, Mallory. Do all of you see it? There!" Brian pointed.

The International Space Station moved silently across the sky and then it was gone.

"Okay, show's over!" Mallory's Mom announced. "Bedtime."

"Aw, Mom," both Mallory and Melody complained.

"Don't forget, you wanted to see the sun rise tomorrow from the other side of the water where it set this evening," Brian said.

"And I need to get a good night's sleep to help my ankle to heal," Grandpa Mike said.

# Chapter 5
## Another Day on Rock Island

Grandpa Mike woke up early, almost an hour before sunrise. He removed and neatly folded the sheets and pillowcase from his cot. He was surprised that his ankle felt fine and that the swelling had gone down overnight. He walked outside and watched as the sky got lighter and lighter.

Brian was the next person to go outdoors. He was very surprised to see Grandpa Mike walking around, without even a hint of a limp.

"Gorgeous morning, isn't it?"

Grandpa Mike turned around. "Oh, I didn't expect anyone to be up yet, Brian. Yes, it is a gorgeous morning."

"I'm happy to see you walking around. Your ankle seems to be a whole lot better this morning," Brian said.

"It must be your clean air," Grandpa Mike answered.

"Or it might be your granddaughter's special touch," Brian said.

"It could be," Grandpa Mike said. "I felt the heat and swelling going down the moment she put her hands around my ankle."

"I smell bacon and eggs. Let's tell Louise the good news while we help set the table," Brian suggested.

Melody and Mallory had already set the table for breakfast when Grandpa Mike and Brian walked in. "Hi, Grandpa," they both said.

"Good news," Brian announced to everyone. "Grandpa Maloney's walking just fine."

"That's wonderful, Dad. You should take it nice and slow today, though. You don't want to risk another sprain," said the twins' Mom.

"That's a good idea," said their Dad. "We can walk slowly back to the ferry dock the way we came. It's much shorter that way."

Louise looked at the brightening sky. "It's almost sunrise. I'll keep our breakfast warm while we all go out and greet the new day."

"And check the compass to see where the sun rises," Melody said.

The sun was just peeking over the horizon, on the opposite side of the water (almost) from where it had set the night before.

"This is something you don't often see from land," Grandpa Mike said. "Back home, we only see the sun rise in the east over Lake Michigan. It sets in the west over land and buildings."

Both Melody and Mallory checked the compass as the sun rose. When they both agreed on the compass reading, they were very surprised to see that the sun wasn't exactly where they had guessed it would rise. They marked the sheet and showed it to everyone.

"The sunrise is a little bit closer to the north than the sunset was yesterday," Mallory complained.

"We checked it again and again," Melody protested.

The sun had risen a bit more by then. Mallory realized any more measurements would be even farther away. "Is it time for breakfast yet?" he asked.

"How does he stay so skinny, Mrs. Maloney?" Louise asked.

"So far so good. I suspect he'll be fine until he stops growing. Then he'll have to watch what and how much he eats," said Mallory's Mom.

"Just like the rest of us," said Mallory's Dad.

"Why is the measurement different from sunset to sunrise?" Melody asked.

"I'm not sure," Louise said. "It might have something to do with the tilt of the earth. I'll bring out another sheet that shows some other times of year and we can compare them after breakfast."

Sure enough. The longest day's sunrise was slightly higher that the same day's sunset. The same was true for the two equinox days and the shortest day. Louise then marked the sheet with the settings that Mallory and Melody had taken.

During his days as a pilot, Grandpa Mike learned a lot about sunrise and sunset angles. He realized that the lessons would have to wait for another day, because Louise and Brian needed to get ready for work. "I'd like to sing a short song that I composed for you. May I?" he asked.

"Sure," Brian said.

"Shall I bring our guitars?" Louise asked.

"Perfect! Can you play the tune *When Irish Eyes Are Smiling*?" Louise and Brian both nodded and went to bring their guitars. This is what Grandpa Mike sang.

*Our Irish eyes are smiling; our thanks for all you've done.*
*Well more than just your duty, we've all had lots of fun.*
*It's time for us to leave now, but new friends we won't forget,*
*We're sending you all good wishes, we are sure glad we met.*

Everyone clapped their hands while Rufus barked his approval.

The twins' Mom looked at her watch. "Thank you for everything, Louise and Brian. We'll need to be on our way so you can prepare for your first tour of the day. We have each others' contact information. Let's stay in touch."

"Indeed we will," said Grandpa Mike. "If you would be so kind as to open the gift shop a little early this morning, I'd like to outfit us all with Rock Island T-shirts as we hike the long way around the trail, back to the ferry dock."

"Are you sure, Grandpa?" Melody asked.

"I'm certain, *Little Miss Healer*. Thanks to your magic touch, my ankle feels fine and even the swelling has gone down."

"If that's the case, would one of you please let the ranger know that we'll be stopping by later than expected?" Melody's Mom asked.

"Of course," Brian said. "Good news like this is always welcome."

Grandpa Mike's plan was to send a donation to the Friends of Rock Island State Park after they got back home. He would also write a letter of appreciation to the Park's Board officers for the help that Louise and Brian provided.

Grandpa Mike told his family when they left the lighthouse, "It's important to write letters of appreciation to those in charge, letting them know how well their employees...or volunteers, in this case...do their work. I'll also email copies to Louise and Brian. Young people like them don't have a lot of work history, so letters like this could be really important in helping them start their careers."

The family continued on their walk along the loop trail around Rock Island. They passed the scenic overlook where Grandpa Mike had tripped over a rock poking out of the ground.

"There it is!" Mallory shouted. "Bad rock!"

*Good rock*, Grandpa Mike thought. *If it weren't for that rock, I wouldn't have discovered Melody's special healing powers.* He decided not to say anything more about that discovery just yet.

As the family walked along the wooded trail on the Lake Michigan side of Rock Island, Rufus had great fun barking at the squirrels scampering on the ground and in the trees. Hawks soaring in the air reminded Melody of the frightened little squirrel she comforted at the Indiana Dunes.

The stone water tower made a nice background for the Maloney family's break for lunch. Two hikers had the same idea. They were on their way back to the campground.

"Let's walk together," Melody suggested. "Maybe you can tell us about some of the things we'll pass on the way."

At the rutabaga field, the hikers explained that the soil was so poor that the early settlers on Rock Island decided that the only crop they could grow was rutabaga.

Mallory was happy to learn about a food he hadn't tried before. "Mom, can we make rutabaga when we get home?" he asked.

"Certainly, Mal. Rutabaga has a fairly earthy flavor, so we can try it in a soup along with other vegetables first," his Mom said.

The campers pointed the way to the Blueberry Trail. "We need to go back to the campground from here, because we're meeting some friends there," one of them explained.

Everyone enjoyed the coolness of the Blueberry Trail before making their promised stop at the island's ranger office.

At the ranger office, Grandpa Mike reported that his ankle healed overnight and that he was able to walk without any problems.

The ranger smiled. "Just be sure to watch out for sneaky rocks popping out of the ground when you least expect them."

Grandpa Mike and the ranger spoke for a few minutes after the rest of the family left the ranger office.

It was almost time to catch the last ferry back to Washington Island, so the Maloneys only had a few minutes to visit the large stone boathouse. Chester Thordarson, an Icelandic immigrant, built the boathouse to resemble the parliament building in Reykjavik, Iceland. The chandelier of horns in the Great Hall drew everyone's eyes upward. Curved windows and carved wooden furniture gave a hint of the grand vision that its builder had for the entire island.

"Too bad we have to leave," Mallory said. "I'd like to learn more about Mr. Thordarson. He must have been a really good explorer!"

"It's always good to dream of coming back to a place you've discovered and enjoyed," said his Mom.

The last ferry of the day arrived, and left on time, with all the Maloneys aboard.


### A Less-Calm Ferry Ride

The wind had picked up somewhat by the time the human passengers and one dog boarded the ferry. Small waves seemed to appear from nowhere once the little ferry left the safety of the harbor.

As the calm water turned choppier, the adults began to think that perhaps their picnic lunch wasn't quite as well digested by then as they might have hoped. Rufus and the children seemed to have no problem; *cast iron stomachs*, as Grandpa Mike put it.

Mallory and Melody thought the ferry's benches were places of confinement, not safety or convenience. Not that their pouting made any difference to their parents. Both Mallory and Melody wound up sitting between two adults. And Rufus? Well, he would have leaped over the ferry's high barrier and right into the water if Grandpa Mike hadn't positioned him between his legs and held a tight grip on his leash.

## Riding the Waves

Mallory was having great fun as the ferry bounced around in the waves. He moved and swayed with each watery hill and valley. Rufus seemed to be enjoying himself as well. The other passengers held tightly onto the benches where they sat.

Mallory noticed an unusually high wave rolling head on toward the ferry and decided, in the instant before it hit, that it would be even more fun to stand up to "ride the wave." The wave heaved up the ferry as Mal jumped up. He was momentarily suspended in midair as the ferry quickly sunk in the depression that followed the wave.

*Uh oh*, he thought, as his feet landed on the ferry's wet deck, heels first. Then fwoomp!

It happened so quickly that Grandpa Mike and his Dad didn't have time to react to the gap that was now between them on the bench. But Rufus was alerted by Mallory's sudden movement. He leaped out of Grandpa Mike's grasp and caught Mallory by the sleeve of his shirt as the ferry pitched forward. Dog and boy slid and crashed into the bow before sliding back as the ferry encountered yet another wave. Rufus tried scrambling to his feet, but all four paws seemed to want to go in different directions. Mallory was annoyed with himself that he always managed to land on his back instead of his feet.

Mallory's Dad and Grandpa Mike were ready to get up and help Mallory back to the bench when his Mom said, "Don't! He's not hurt and he knows better than to disobey us. What if this ferry had an open bow like some other ferries? He and Rufus would have both been washed overboard and drowned. **Good boy**, Rufus!"

"**Bad boy**, Mallory Joseph Maloney," she added.

*Gee, mom doesn't care if I'm hurt or not*, Mallory thought. *She's just mad because I didn't want to sit on those dumb old benches. And she doesn't yell at Mel as much, either.*

Mallory was thinking all those thoughts, and more, about the poor state of mothering he had to endure.

"And there will be no dessert for you this evening," Mallory's Mom scolded.

That just reinforced his thoughts. *I don't get to have ANY fun*, Mallory continued thinking as he finally managed to stand upright.

Fortunately for Mallory, the ferry was nearly at the dock. The landing cut short any further scolding. Everyone was a little damp from the waves, but otherwise fine after the short trip.

As they were leaving, Grandpa Mike asked the captain how often the waves got that high.

"Oh, once in awhile, but this was a bit rougher than usual for this time of year," the captain said. "The waves were pretty exciting, weren't they, young man?" the captain asked Mallory.

Mallory looked down and didn't say anything, thinking that a ferry's captain probably needs to see everything on his boat. He hoped he wasn't going to get another scolding.

*Whew!* Mallory breathed a sigh of relief when Grandpa Mike asked the captain for directions to the local grocery store.

"Before we do anything else, let's go back to the inn, shower, and change into fresh clothes," Mallory's Mom said.

"Good idea, Mom." Melody didn't like to wear the same outfit two days in a row, even if no one else would notice...or even care.

"I need to make a stop at the grocery first," Grandpa Mike said.

As they drove to the grocery store, Grandpa Mike explained that he was very grateful for all the kindness and helpfulness that Louise and Brian had shown to all of them.

"All four docents are going to enjoy a pizza party tonight."

That's all Mallory needed to hear. "Neat Grandpa! Can we go back for the pizza party?"

"I don't think so, Mal. This is a party for the docents. Louise and Brian shared what food they had with us, and the other two docents gave up their bedrooms for us. I phoned the ranger last night to make arrangements. With dinner last night and breakfast this morning, we also 'ate them out of house and home,' as the saying goes. That includes the dog food for Rufus. The docents picked up everything on my shopping list this morning and went back to the lighthouse so they could continue doing their work giving tours. They now also have a good supply of extra granola bars and other treats."

"So that's what you were talking about with the ranger," Melody said.

"That's very thoughtful, Dad," said Melody's Mom, and her Dad agreed.

"Here we are," Grandpa Mike said as they parked in front of the grocery. "Why don't the rest of you look around and see if there's anything you'd like to pick up? Meanwhile, I'll square up with the store manager."

# Chapter 6
## Back on Washington Island

"We heard you were unexpectedly delayed on Rock Island," Ms. Karen said as the family entered the inn's lobby. "I hope everything is alright now."

"Yes, thank you Karen," Grandpa Mike said. "A small rock was my downfall."

Mallory disagreed. "But you didn't fall down, Grandpa."

"That's very clever, Mallory. Your Mom's fascination with words seems to have rubbed off on you," Grandpa Mike said as he patted Mallory's head.

"Grandpa's ankle is all better now," Melody said.

"Indeed it is," Grandpa Mike agreed. "Nurse Nightingale's cool hands absorbed all the heat from my swollen ankle and it healed overnight."

"Who is Nurse Nightingale?" Melody asked.

Ms. Karen answered her, "Florence Nightingale was an English nurse who lived in the 1800s. She saved many lives by improving health care conditions, especially among the poor. If I'm not mistaken, she collected statistics when treating wounded soldiers to prove how important it was to keep germs from spreading disease."

"That's neat!" Melody said. She remembered looking at bacteria through a microscope at Yellowstone Park. She also liked the name Nightingale. She liked birds, especially their songs.

"Statistics?" Mallory asked.

"It's a branch of mathematics," his Dad said. "You collect lots of facts or numbers and then we use different types of arithmetic to analyze them and come up with an explanation to support your idea or prove it wrong."

"I like collecting things," Mallory said. "I don't know about collecting numbers, though."

"Me, neither," Melody said. She usually wasn't so quick to agree with her brother.

"Well, now. I think it's time to get into some fresh clothes and start thinking about dinner," Grandpa Mike said.

"You'll probably want to try different places on Washington Island. What kind of food might interest you this evening?" Ms. Karen asked.

"I'd like to go someplace that has really good desserts," Melody teased, looking at her brother.

"Don't get uppity, young lady, or you'll be missing dessert as well," her Mom said.

The restaurant Ms. Karen suggested was close to the inn, near the airport. Lunch on Rock Island seemed like it was a long time ago.

The food was delicious. "Locals always seem to know the best places to eat. It's nice that Karen didn't mind recommending a competitor," Grandpa Mike said.

Everyone, except Mallory, enjoyed the restaurant's pastries for dessert. He didn't mind so much, though, because he was listening intently to the conversation of some divers who were seated at the table next to theirs. The group was discussing the highlights of the day's dive. They had been exploring the shipwrecks sunk beneath the mildly choppy waters surrounding Washington Island.

Visions of pirates and treasure occupied Mallory's thoughts, even though none of the divers mentioned pirates or treasure.

After dinner, Grandpa Mike said, "It's still light out. Would anyone like to see the *Stavkirke* nearby?"

"What's that, Grandpa?" Mallory asked.

"I understand that it's a replica of Norwegian wooden churches from medieval times. These buildings were supported by posts, or staves. That's why they are called *Stavkirke,* which means Stave church. Their roofs have lots of layers and levels; it is a type of architecture that is quite rare."

"That sounds really interesting, Dad," Mallory's Mom said. "I understand there were a lot of Norwegians who settled here, along with the Icelanders we learned about at Rock Island."

"I've always been fascinated by the Scandinavian people, especially Norwegians, and their culture," Grandpa Mike said. "I'd sure like to see an example of their building design and construction, especially since it's nearby. I understand that there are a lot of Norwegians living in Wisconsin. Fascinating people, Norwegians."

"Have you ever been to Norway, Grandpa?" Mallory asked.

"No, I haven't, but I would like to visit Norway someday. I'd like to see the fjords, mountains, the Northern Lights and Midnight Sun."

"Northern Lights? Midnight Sun? That sounds really interesting, Grandpa," Mallory said. "I want to go to Norway!"

"I suspect that someday you will, Mal," Grandpa Mike said. "I've heard that sometimes the Northern Lights are visible from Washington Island during the colder months."

The *Stavkirke* building itself was closed, but the outside was like nothing anyone had ever seen before. They walked around the building to admire it from all different angles.

Mallory didn't want to leave. "Thanks for letting us know about this, Grandpa." What Mallory really wanted to do was climb up to the first level roof and look down. He knew better than to even hint at what was in his mind. He didn't want to miss another dessert!

Melody got out her notepad and started sketching what she saw.

"I missed a lot of it and I think I got some things wrong, but I wanted to try to remember how it looks," Melody explained.

"This has been quite a day," Melody's Dad said. "I think it's time for all of us to get a good night's sleep. We can explore more tomorrow."

Just then Melody's Mom's cell phone rang. It was Bobby's Mom. She asked if the Maloneys would like to join their family tomorrow evening for a game they liked to play on Washington Island. "It's a family tradition," she told Melody's Mom.

Melody's Mom listened for awhile, nodding her head every once in awhile. "Um, hm. Yes, that sounds like fun. Thank you. We'll see you tomorrow evening, then."

"What, Mom?" Mallory wanted to know.

"What, Mom?" Melody wanted to know, too.

"It's a surprise. If I tell you, it won't be a surprise, will it?"


## A New Day

The young woman greeted the Maloneys when they entered the Welcome Center the next day. "Welcome to Washington Island. Is this your first visit here?"

Mallory hurried to tell her, "We already went to two beaches and saw the lighthouse on the other island and lots of other stuff."

Mallory's Mom explained, "Most people stop at a place's welcome center first. We were so excited about exploring that we've been running around quite a bit. And yes, as my son said, we've already been to Rock Island. What do you recommend for us to visit here on Washington Island that would be at a slightly more leisurely pace?"

Ms. Margaret got out a map of Washington Island. "Have you walked our short nature trail that's right here, next to our building? How many more days will you be here?"

Mallory's Dad said, "We're planning on catching one of the early ferries back to the peninsula the day after tomorrow."

"If you mark on this map where you've been, I'll be happy to make some recommendations for you," Ms. Margaret said.

Mallory's Mom put a dot near the areas they had already visited.

Ms. Margaret brought out another map. "If you're looking to be out in nature, here is a guide to Washington Island Nature Preserves of the Door County Land Trust.

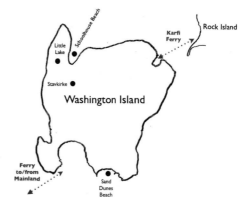

You've already visited the Little Lake Nature Preserve. We also have a bird sanctuary, marshes, swamps and harbor areas. You could also visit the Historic Island Dairy and the lavender fields that surround the Dairy.

Ms. Margaret continued, "We have several museums where you can learn a lot of this island's history. Two of them are on Jackson Harbor Road. You might have passed them both on your way to the Karfi Ferry to Rock Island. One is the Farm Museum and the other is the Maritime Museum."

"We visited my aunt and uncle's farm in Illinois, Ms. Margaret," Melody said. "I think the Maritime Museum would be more interesting. That's about ships, right?"

Mallory was getting bored with the talk about museums, but he perked up with his sister's word *ships*. "And Death's Door, right?"

"That's right, young man," Ms. Margaret said.

"Thank you. You've been very helpful, Margaret. Now we need to figure out which areas we want to visit in these next two days," Grandpa Mike said.

"Let's walk along the nature path and talk about what we'd like to see," their Dad suggested.

"Great idea, Mort," their Mom agreed.

"Lavender and birds," Melody announced, as they walked along the Nature Trail.

"Swamp, dairy and Maritime Museum," Mallory quickly added.

"No fair, that's three!" Melody yelled.

"Suppose we sit down on this bench. We can then look at the map and plot our strategy," Grandpa Mike said.

"Do you know what day it will be tomorrow?"

Mallory was confused. "Um, tomorrow's Friday, Dad."

"And what happens on Fridays here?"

Melody's eyes lit up. She poked her brother. "It's fish boil, silly. I thought you knew everything about food."

"Let's mark your five places on the map," their Mom said. She pointed to two of the places that were farther away (both Mallory's selections), and the other three that were closer to the inn.

"Let's work backwards," their Dad said. "We need to be back at the inn in time for the fish boil, and that means changing into clean clothes before we eat, right?" said their Dad, expecting agreement from his children.

Both Melody and Mallory nodded.

"Then, it probably makes sense for us to visit the two farther choices today, and the three closer-in choices tomorrow. How does that sound to all of you?"

"No fair! Both those places are what Mal picked," Melody said, pouting.

"Don't worry, little Miss Nightingale," Grandpa Mike said, winking at her. "I'm sure you'll find plenty of fun waiting for you today." Her Mom had already told Grandpa Mike about the surprise.

Mallory could have spent the entire day at the Maritime Museum. The two converted fishing sheds and the outdoor displays of life on the shores of Washington Island transported him into another world.

Mallory wondered what it felt like to be aboard a fishing or trading boat or ship. He wondered how hard it was for sailors to use the stars to find their way over water that didn't have any landmarks. He wondered how sailors and fishermen managed to keep their boats safe in dangerous storms, and what it felt like when the storm was too strong and caused the boats to sink.

"Wake up, Mal. It's time for lunch." His Mom's voice tunneled through his daydreams. Until then, Mallory didn't even realize he was hungry.

After lunch, Mallory's Dad said, "Let's see how different the wetlands are like here, compared to Indiana Dunes."

"Neat idea, Dad," Mallory said.

"Melody, do you remember our visit to the bog at Indiana Dunes?" Melody's Mom asked.

Melody nodded.

"Wetlands are an important part of our natural environment," Grandpa Mike said. "That's why local and state governments, and even our national government, have created programs for protecting them. It will be interesting to compare this area with what you saw in Indiana."

Rufus led the way into the Coffey Swamp natural area from where they parked on the side of the road. Dogs seem to have an instinct for finding natural areas.

Muddy soil surrounded a shallow pond.

The twins' Mom pointed to the water and the splotchy, spongy-looking ground around it. "It looks like it's dried up quite a bit; it's already late in the summer, and we've not gotten much rain lately."

"Look Mom," Melody said. "That looks like one of those insect-eating plants we saw at the bog at the dunes. There's no moss floating on this pond, though. What was that moss called? I forgot."

"That was *sphagnum* moss covering the bog, Mel. And that plant looks like a pitcher plant. Mal, that's the plant your sister teased you about," their Mom said.

"Oh yeah, I remember. She called it a boy-eating plant. Only it eats bugs, right?"

"That's right, Mal. It eats insects," his Mom said.

They came to a swampy forest. "Dad, it's almost like what we saw while Mom and Mel were at the bog," Mallory said.

"Well, well. I guess I am seeing part of your trip to the Indiana Dunes without even going there," Grandpa Mike said. "Even better, I escaped having to climb those hot dunes!"

Melody and Mallory giggled. "And you escaped finding buried treasure," Mallory added.

## A Fun Way to Hunt Deer

After a light supper, the twins' Mom announced that the whole family was going deer hunting.

Melody was speechless. No one talked about shooting animals. That didn't sound like fun at all to her.

Mallory wasn't exactly pleased either. "Is this the surprise you promised?" he asked.

"It is," said their Mom. "Except, we're not shooting deer. We're going deer counting. Bobby's Mom invited us. Two teams. Their family and ours. We'll drive around a wooded area where deer like to come out when the sun goes down. The idea is to count the deer we see and mark down the initials of who saw the deer. At least two people need to see the same deer to gain one point. Whichever team loses gets to buy everyone ice cream at the end of the game. After all, counting deer is hard work!"

Melody and Mallory were both very relieved.

"Oh, good," Melody said. "If we take Rufus, we're going to win. He'll bark and let us know there's a deer somewhere."

"If we see more than one deer at a time, do we get points for all of them?" Mallory asked. He wanted to make sure he understood the game's rules. He was also happy that even the losing team got to eat ice cream.

"I believe so," his Dad said. "We'll ask to make sure."

So off they went to meet up with Bobby's family. Everyone agreed that it was okay for their larger group to compete with the smaller Maloney group.

"Rufus counts for at least four people," Bobby said. "Everybody know that dogs can smell deer a mile away."

And so the game began.

The game would end a half hour later at the ice cream shop, where the two families would trade their score sheets and add up the other family's scores. Both families believed in the honor system and trusted that no one would add extra points to their sheets that they did not earn.

At the ice cream shop, 30 minutes later, Grandpa Mike took his time adding up the number of deer that Bobby's family saw. He then passed the sheet to Mallory, who would confirm the count.

"Bobby's family: 97," Mallory read.

"Mal and Mel's family: 92," Bobby read. "We win!"

Grandpa Mike smiled. This was exactly the outcome he had hoped for. He turned over the counting sheet and passed it to Bobby. "Well done, team. Please write down the kind of ice cream you would like to order, and what size."

"Thank you, Grandpa Maloney," Bobby said. "I think we just won because we've been playing this game for lots of years, and this is your team's first time. You guys saw a lot more deer than we did our first time. I think we only saw about 25." He then gave Mallory the other counting sheet for the Maloneys to write down their ice cream choices.

Melody asked Bobby, "Why don't we see so many deer during the day? Do they only come out when the sun is setting?"

Bobby answered, "I'm not sure, Mel. I think it has something to do with their eating habits. Does anyone know?"

One of the servers heard Bobby's question when she was bringing ice cream cones to another table. "My aunt is an animal biologist. She says that deer mostly come out during dawn and dusk. That's when they feed. She says that deer have good night vision and sometimes roam around at night, so that's why you have to be careful when you drive on dark roads. During late summer, like now, lots of flies are around. The deer try to escape their bites by by leaving the woods for open areas, like roads."

"Thank you, Mindy," Bobby's Mom said. "We appreciate your sharing your aunt's knowledge with us."

"Mom, that was a lot of fun," Mallory said on the way back to the inn. "We should try counting things on the way back home. Maybe cows?"

# Chapter 7
## Fish Boils, Boiling Chips, ... Extremophiles?

**The Last Explorations on Washington Island**

"Today we're going to the places I picked, right?" Melody asked the next morning.

"And one of mine. Don't forget about the dairy," Mallory said.

"Let's not forget about the fish boil tonight," Mallory's Dad reminded both children. "It will be the grand finale of our visit to two of Door County's islands."

Their first stop was the bird sanctuary.

"According to this pamphlet, there are more than 200 species of birds that have been recorded on Washington Island," said Mallory's Mom. "The donation of this preserve was the beginning of Door County's Land Trust on Washington Island in 1997. The name of this preserve, Domer-Neff, is dedicated to the two women who loved this area and worked hard to make it a peaceful, welcoming place for birds and other animals to make their homes here."

Melody liked the sounds of many different kinds of birds that followed the family's walk along the loop path of the bird sanctuary. They saw and heard blue jays, red-wing blackbirds, and even some squawking crows.

Melody's Mom read more from the pamphlet. "It says that school children planted more than 500 trees and bushes that are native to this region to attract animals that would find it a perfect place to call home."

"What a fine example of people and their local government working together to preserve, and even improve, the environment," Melody's Dad said.

"That's why it's called a *preserve*, Dad, right?" Melody asked.

"I like peach preserves," Mallory announced.

**51**

"Say, aren't we close to the Stavkirke that we visited the other evening?" Grandpa Mike asked.

"We are, Dad," Mallory's Mom said. "We could take a better look at the building and walk the trail around it if you like."

"We should go back there. I liked all the roofs. Then we could go to the Dairy, right?" Mallory asked. He thought flat trails were kind of boring, even if the songs of different birds were fun to listen to.

"Let's see if we have time," Grandpa Mike said. "It's better to enjoy visiting new places. Besides, there may be other similar buildings on the way back through Door County's mainland."

"Hooray, we're at the Dairy!" Mallory was the first to announce as the large stone building came into view.

"Hooray, we're in lavender land!" Melody shouted at the sight of all the lavender plants surrounding the Dairy.

"It looks like you're both right," said their Dad. "Shall we take a look at the building first, and then enjoy the lavender fields in all their outdoor glory?"

Rufus barked his disapproval at being left in the car as everyone else got to explore the building known as the Historic Island Dairy.

After walking around the art gallery and history exhibit, they visited the lavender shop. Melody wanted to get the lavender dog bath for Rufus "so he'll be the best-smelling dog we know," she said.

Mallory wanted the cherry lavender jam as a tasty souvenir, and his Mom thought the lavender balsamic vinegar would be nice to try.

Mallory's Dad and Grandpa Mike were more interested in learning about the history of growing lavender on Washington Island. They learned that potatoes were once an important crop grown on Washington Island and were exported in large quantities all around the Great Lakes region. The potato crops are now mostly gone. Also gone are most of the cattle and cheese-making activities.

The lavender fields were started from seeds brought from France. *Lavender is a nicer reflection of Washington Island's French history than Door County's unfortunate name, which came from the French for Death's Door*, Grandpa Mike thought.

"It looks like we won't have time to re-visit the Stavkirke. It's time to start heading back for the fish boil," the twins' Mom announced.

No one presented any objection.

## Watched Pots DO Boil

Two of the inn's "boilers" had just finished filling a big black kettle with water when the Maloney family entered the courtyard at the back of the inn. A stack of firewood had been set beneath the kettle.

"That's a HUGE pot!" Mallory exclaimed. "That's not all for us, is it?"

"No, I don't think we could eat all of what's going in the pot," his Dad said. "More people will be coming. It's good we arrived here early. You and Mel picked the best picnic bench for us to watch the fish boil. We've got a perfect view of the preparation from here."

"Absolutely. Ringside seats," said Grandpa Mike.

One of the boilers lit the firewood and soon the water in the pot started to boil. The other boiler took a sack and poured a good amount of its contents into the boiling water.

"Is that salt?" the twins' Mom asked. It looked like far too much for the size of the kettle.

"It sure is salt," answered the boiler. "It's only half of what we'll use for tonight's boil."

The twins' Dad said, "Adding salt would certainly raise the density of the water. We chemists like to use the fancier term *specific gravity*, but the word density will work just fine." He didn't want to seem like a know-it-all, so he then asked, "The water's higher density would make all the ingredients float to the top instead of sinking to the bottom of the kettle, right?"

"Hmm. That makes sense," said the boiler. "Let's ask the master boiler."

"You're absolutely correct, sir," said the master boiler. "The salt also causes the oils from the fish to foam up...and out...when the water boils over. This way the extra oil winds up on the ground and the fish tastes less *fishy*. You'll see what I mean in a few minutes."

"Dad, if the salt makes the water *more* dense, does that mean that the fish oil is *less* dense?" Mallory asked.

"That's right, Mal. Oils are less dense than water. That's why you have to shake an oil-and-vinegar bottle of salad dressing to mix it up and then pour it quickly before the oil rises to the top again," his Mom explained. "And before you ask, vinegar is mostly water, so its density will be just about the same as water."

The first boiler brought out a perforated metal basket filled with potatoes and onions. The boiler waited until the water came to a full boil before lowering the basket deep into the water. Then he went in and brought out a smaller metal basket filled with fish. The master boiler added more salt and then nodded to his assistant to lower that basket into the water.

The master boiler watched the kettle closely, once the water started boiling again.

"I thought a watched pot never boils," said Grandpa Mike, as the twins' parents groaned at his joke.

"Dad, I'm sure they hear that at every dinner," the twins' Dad sighed.

"Actually, it's quite rare, sir. People are so fascinated with the boil, they forget those old sayings," said the master boiler.

*A good example of knowing how to treat your customers right,* Grandpa Mike thought.

"Speaking of fascinating, now's the time for some fireworks to finish the cooking," the master boiler announced. He took some kerosene and threw it on the fire. Flames shot up and surrounded the pot, causing water and the foamy fish oil to "overboil." The overboil spilled out over all sides of the kettle and onto the fire, putting it out.

"Wow!" both twins exclaimed, along with some of the other guests. Others, who had been to fish boils before, had their cameras ready and snapped pictures of the flames.

## Another Extremophile in the Family?

As the dinner plates were being prepared, the twins' Dad shook his head and smiled.

"You look like the cat that caught the canary," said the twins' Mom.

"The overboil reminds me of a chemistry lab I took in college," their Dad began.

"I think this is an outdoor *cooking* class, dear, not a lab experiment," their Mom said teasingly.

"Actually, it WAS a chemistry lab class that just came to mind," their Dad answered. "Chemistry wasn't my best subject back then."

Melody and Mallory knew from their Dad's smile and the tone of his voice that they were about to hear a funny story.

Their Dad continued, "I was reading the instructions for a laboratory experiment right there for the first time, at the lab bench, so I didn't have a good overall idea about what the experiment was all about. We were supposed to be making methyl orange. Methyl orange is used as a pH indicator. Mal, Mel, do you remember Dr. Ethyl's explanation about pH?"

The twins thought for a moment. It was several months since they thought about pH. Then Melody said, "I drew a see-saw with different colors. It had water and milk in the middle..."

"And colas and vinegar were on the acid side of the see-saw," Mallory interrupted his sister.

Grandpa Mike congratulated his grandchildren. "Excellent, both of you! Drawing pictures of things helps you to remember them, doesn't it?"

Then their Dad said, "Methyl orange is a beautiful orange dye. It changes its color from orange to either red or yellow, and that color tells you the pH of the acid solution you're measuring."

"So it's like using orange juice instead of paper to find out the pH of something," Mallory said.

"That's right, Mal, except that methyl orange is clear, more like apple juice," his Dad said.

"Better not drink it," Melody giggled.

Their Dad normally didn't talk much about himself. This time was different, though. It seemed to everyone that he was really enjoying remembering this story and sharing it with his family.

"There was this really cute girl in my lab class that I was trying to impress. I pretended that I studied the lesson, so right away I started weighing out the chemicals listed for this experiment. Then I began following the directions, step by step without really knowing what I was doing. By the way, this style of conducting experiments is called "cookbook chemistry."

"Cute girl, huh?" the twins' Mom asked. "What was her name?"

"I don't remember, Agnes. Anyway, I DID impress her, except not in the way I wanted to," their Dad admitted.

"What happened, Dad?" Both Melody and Mallory were anxious to hear what happened next.

"After I added the chemicals in the flask and started heating it, I noticed that I forgot one piece of instruction. It said, *add boiling chips to your flask for safer heating and to prevent it from bubbling over.* By this time the liquid was quite hot, so I thought I'd better throw in some of the boiling chips right away."

"Ah hah!" the twins' Mom exclaimed, laughing heartily. "I'll bet you impressed her, all right. Just how orange did you get?" Their Mom obviously knew a little something about "kitchen chemistry."

Melody and Mallory were confused. So was Grandpa Mike. "What's so funny?" they asked.

"There was an eruption! It looked like a volcano spitting out hot lava, except my lava was orange," their Dad said. He stood up and pantomimed a volcanic eruption with his hands. "That's what happens when you put *cold* boiling chips into a *hot* liquid. Those boiling chips do exactly the opposite of what they're supposed to do when you put them in BEFORE you start heating up the liquid. Our magnificent show here reminded me of that awful day."

"What about the girl?" the twins' Mom asked. She wondered what kind of person her husband was interested in back during his college days.

"These were the first, and *only*, words she ever spoke to me: 'Forsooth! Thine flask bubbl-eth over well, methinks.' She and the class were all laughing really hard, both at me and at the eruption. I was covered with orange splotches and my hands were orange nearly all over. Good thing I was wearing a lab coat and safety goggles."

"You never told me about that, Mort," Grandpa Mike said, pretending to scold his son. "Hot liquid! You must have gotten blisters from burning yourself."

"No, that's the amazing thing, Dad. Even though I was really close to the flask, I didn't get burned at all. In fact, I don't recall it feeling particularly hot, even on my hands, and they were covered all over in boiling hot orange dye."

Mallory and Melody looked at each other. They recognized a certain similarity with Melody's experience at Old Faithful geyser. "Extremophile. Do you think Dad's an extremophile too, Mel?"

"What are you two whispering about, Mal?" their Mom asked.

"Oh, nothing Mom. I just thought about how funny Dad probably looked," Mallory answered.

"I take it you raced to wash your face and hands before the dye set," said his Mom, turning her attention back to her husband.

"Oh, I did. I was *so* embarrassed; my face was almost as bright as the methyl orange. That was my last attempt to impress that girl, though. As a matter of fact, I don't even recall her name. Barbara, I think it was. Or maybe Nancy," the twins' Dad replied.

"Likely story," laughed their Mom. "It was probably Susan, or Elizabeth."

Melody was happy to see a new and different side to her father.

"That was a funny story, Dad," Melody said. "It makes this fish boil really special."

"And your story makes the fish taste even better," Mallory said.

Everyone agreed.

# Chapter 8
## Crossing Death's Door a Second Time

"This has been a great vacation, Dad," the twins' Mom said to Grandpa Mike as the family got into their car. They were near the front of the line of cars waiting for the ferry to take them back to the mainland.

"Indeed it has," Grandpa Mike answered. "I learned something about my son that I never knew before."

"We learned that Dad is really funny," Melody said to her Grandpa.

"What, you didn't know I'm funny? Even Rufus knows I'm funny, right, Rufus?"

Rufus barked when he heard his name.

"See? Rufus knows," the twins' Dad said.

*Ah, this vacation has been a good thing for Mort*, their Mom thought. *That was a very funny story he told us. You think you know someone well, but you learn something new all the time.*

The ferry's arrival interrupted their Mom's thoughts. "Everyone ready to drive off into the sunrise?" she asked.

"That's funny, Mom. The movies always show people riding off into the sunset, not sunrise," Melody said.


**Weather Reporting**

Melody and Mallory were anxious to get out of the car and watch as the ferry moved smoothly through the water. They were starting to miss Washington and Rock Islands already.

Then they both saw it at the same time.

"Look!" Melody said.

"What's that?" Mallory asked.

"That's the Washington Island buoy," one of the passengers said.

"Boo-ee?" Melody asked.

"That's right, but it's not spelled the way you think," said the passenger. "It's spelled b-u-o-y."

"Please tell us about it," Grandpa Mike said. "My grandchildren, and the rest of us, would like to learn more about buoys. We're the Maloneys," he said and introduced everyone.

"I'm Kate and a member of the NOAA team that monitors our buoys in the area."

The Maloneys weren't the only ones interested in the strange object floating in the water. Many of the other passengers nearby gathered around as Ms. Kate started talking about the buoy.

"First of all, NOAA, which sounds like the Noah who built an ark a long time ago, stands for *National Oceanic and Atmospheric Administration.* NOAA is a U.S. government agency that does research on the oceans and atmosphere, just as its official name says. This research is used for many purposes. The activity most people associate with NOAA is our warning of dangerous weather conditions," she said.

"That's the information used by weather forecasters, right?" someone asked.

"Yes," Ms. Kate answered. "Our information is often combined with data from other public and private sources."

"Please tell us about the buoy, Ms. Kate," Mallory said, hopefully.

"I'll try to give you an idea about these fascinating objects," Ms. Kate said. "First of all, NOAA buoys measure and send out data on all sorts of things, such as the direction and speed of wind, barometric pressure and wave activity."

"How come they don't float away?" Melody asked.

"They're anchored, aren't they?" another passenger asked.

"Yes, the buoys are moored to the bottom of the lake. Nylon and other materials are used for deeper waters, such as here in the Great Lakes and in deep oceans. Chain is used for shallower waters," Ms.

Kate said. "There are floats and other types of devices used, depending on the depth and conditions of the water. The buoys here on the Great Lakes are taken in each fall because we don't want them damaged by ice. They are serviced and made ready for the next year."

"How do they operate to send out the signals?" Grandpa Mike asked.

"We use solar cells to charge their batteries," Ms. Kate said. "They are marine batteries, made specially for water environments."

"How does the data collected get sent out?" someone else asked.

"Satellites receive the transmitted data and then send it to receiving facilities on land," Ms. Kate said.

"We saw the space station on Rock Island," Mallory announced. "Do they get the signals from the buoy?"

"No, I don't believe they do," Ms. Karen said.

Unfortunately, there was no more time for questions. The ferry was approaching the dock and it was time for everyone to return to their cars.

"It was interesting to learn about how the buoy helps warn about bad weather coming, " Melody said when they got in the car.

"It sure was," her brother agreed. "I even forgot that we were passing Death's Door!"

# Chapter 9
## On the Way Back

The trip going home took much longer than the drive to, and then directly *through*, the peninsula of Door County to catch the ferry to Washington Island. They zigzagged southward as different places caught someone's interest.

"I want to see the goats on the roof," Melody announced.

"I'd like to see them as well," Grandpa Mike agreed. "That's something that none of us will probably see anywhere else."

Mallory wanted to see the caves that were described in one of the brochures, but the thought of Swedish pancakes with strawberries and whipped cream at the restaurant that has goats on the roof won his vote.

There was plenty of time, so these weren't "either/or" kinds of decisions that had to be made. The family enjoyed beaches, waves and caves, nature paths, harbors, lighthouses...

Ah, lighthouses!

Melody overheard a group of people talking about the lighthouse in Sheboygan. They said the lighthouse was at the end of a really long pier, about one-third of a mile into Lake Michigan. Even though the family had visited and climbed to the top of several lighthouses in Door County, she felt she had to see and walk along the long pier leading to the Sheboygan Lighthouse.

"Well, why not?" Grandpa Mike agreed. "It may be a long time until we come back this way."

"What a spectacular view!" Melody's Mom exclaimed when she saw the long breakwater and pier leading to the Sheboygan Breakwater Lighthouse.

"It was worth getting off the highway for this," said her Dad.

"Can we go all the way to the lighthouse?" Melody asked.

"Only if we leave Rufus in the car," Melody's Dad replied. "He'll want to leap into the water as soon as we step onto the pier."

"Nonsense," Grandpa Mike said. "Rufus and I will stay here together on solid land. My eyesight isn't the best, and I don't trust my balance, even though the pier seems wide enough to be safe."

Mallory wasn't so sure that his balance was any better than Grandpa Mike's, but he didn't want to miss this exciting adventure.

The setting sun reflected on Lake Michigan's eastern horizon when parents and children rejoined Grandpa Mike and Rufus.

Mallory said proudly. "Grandpa, that was the best walk I have ever taken! I even walked close to the edge of the water!"

Grandpa Mike smiled, sat down on a piling and took a photo of the long pier to remind Mallory of his grand accomplishment.

Melody didn't know what the big deal was with her brother. She had good  balance, just like her Mom.

# Chapter 10
## Next Adventure: Discussing the "Loch"

"That was a long ride," said Grandpa Mike as they dropped him off at his house. "Thanks for driving, Agnes and Mort. Even though my ankle has healed nicely, it's better off not needing to push down on car pedals for a couple of hours at a time."

"Not at all, Dad," the twins' Mom told him. "We loved having you with us."

"Grandpa, are you coming with us next time for the long picnic at the loch?" Melody asked.

"I believe I will, Mel," Grandpa Mike said. "It's been a long time, too long, since I've last fished. I've been meaning to reacquaint myself with the joys of feeding worms to the fish, now that I've got some time on my hands."

Actually, Grandpa Mike was beginning to feel the burden of too *much* time. Since he sold his software firm to a big company, he felt like he was drifting, not accomplishing anything worthwhile. That's one of the reasons why he decided to fund these mini-expeditions for his grandchildren and their parents.

Grandpa Mike had gotten a call recently from one of the executives at the company that bought his software firm. The manager hoped to hire Grandpa Mike as a consultant to help sites that were still using old computer systems. Many of their newer employees had no experience with "heavy iron" mainframe computers, so Grandpa Mike's experience from his military service, where he first learned Cobol, would be very welcome.

There might not be much time for weekend fishing if he decided to do that kind of consulting. Installing and debugging software has to be done during evenings and weekends.

Grandpa Mike was sorry he didn't take his own son fishing more often. *Mort doesn't have the experience to teach my grandchildren the joys of fishing*, he thought to himself.

A few days later, Grandpa Mike called.

"Interesting news, Mort. Al says there should be plenty of fish in the pond. Actually, he likes to call it the "loch" rather than a pond, even though he's Norwegian and not Scottish," Grandpa Mike said.

"That's strange," the twins' Dad commented.

"Not so strange when you hear the story. I'll tell the story to all of you when you come over to dinner tomorrow night. I'm looking forward to Aggie's bringing all those vegetables I hear you've been harvesting from your garden. They'll go well with my grilled salmon."

Over dinner the next evening, Grandpa Mike explained that his friend's last name was Ness. "I hear that the pond has a huge fish that has been jumping high out of the water at night lately. It's been making a big, noisy splash when it lands back in the water."

"Ah," the twins' Mom said. "That's starting to make sense. Their last name is Ness, they have a pond that has a big, mysterious fish in it, and the Norwegians have stories of lake monsters too. I like the name *Loch Ness* much better than *Pond Ness*. It just doesn't have the same sound to it, does it?"

"Isn't there a Loch Ness monster or something?" Mallory asked.

"That's right," Grandpa Mike said. "The original Loch Ness is in Scotland. For many years, people have reported that they've seen a 'monster' out in the deep waters of Scotland's Loch Ness."

Mallory could hardly wait until the Labor Day weekend at the beginning of September. Three whole days to hunt down a big, monster fish! That sounded like great fun!

"Loch Ness monster. Sounds pretty 'fishy' to me," his Dad said.


The map on the next page shows the location of the next book.

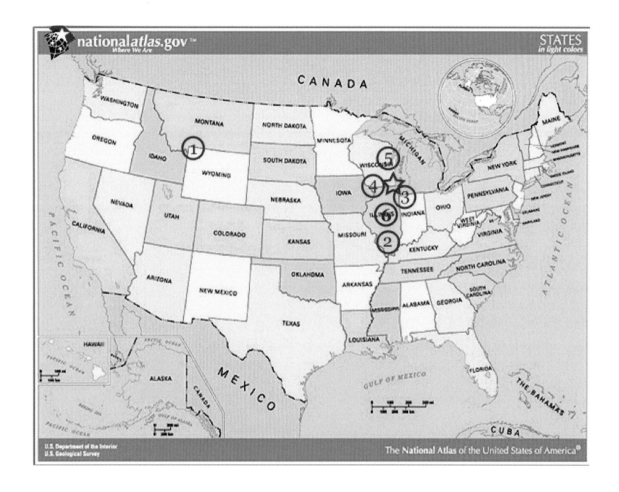

# INDEX

Made in the USA
San Bernardino, CA
02 July 2016